Musings of the Wandering Whisky Whisperer

By

John Clement

"If you love the whisky, the whisky will love you."

Juhani 2016

Cover design by Robin Ludwig Design Inc.
www.gobookcoverdesign.com

Photography from John Clement and
Sean Clement Photography

Dedication

This book is dedicated to Marc Zanutto, a kind and gentle soul, who shared my passion for whisky and gave me encouragement along the way. If there is whisky in heaven, I'm sure you will find Marc at the table, enjoying a wee dram or two.

Praise for Musings of the Wandering Whisky Whisperer

"John Clement's new book, *"Musings of the Wandering Whisky Whisperer"* is a fun-filled look at the joys of whisky through the anecdotes of an aficionado. The perfect whisky cocktail, it combines a base of knowledge, sweetened with personal experiences and spiced with a dash of humour. Easy reading for the beach or the fireside. "Davin de Kergommeaux Author of the award winning book, *Canadian Whisky: The Portable Expert*

"In his whimsical book, John Clement, THE Wandering Whisky Whisperer, takes us on a journey to discover that the true essence of whisky goes beyond its rich, diverse, complexity to reveal secrets about living well and life itself," Edward Wilson, Scottish forester and whisky aficionado.

"Musings of the Wandering Whiskey Whisperer is both amusing and enlightening. If you drink whiskey this will give you a deeper appreciation and if you don't drink whiskey you will after reading this book." Denis O' Callaghan Irish whiskey connoisseur promoter and real live Irishman.

Table of Contents

Acknowledgements 10

Preamble 13

Introduction 15

My Whisky Journey 19

Warning! 25

What is Scotch Whisky? 29

Whisky Maturation Process 31

My Own Maturation Process 37

How to Whisper Whisky 39

Let's Start 43

The Musings of the Wandering Whisky Whisperer 53

Respect 55

Patience 63

Taste the Difference 65

Moments 69

View From the High Road 71

Scotchland 73

Down to Earth 75

Grit 77

Take Note..79

Appreciate the Finer Things in Life..................................81

Whispering Whisky as Mindfulness in Tastings 87

Whisky Makes You Happy!..95

The True Value in a Bottle of Whisky...............................97

Whisky Will Protect You..99

Whisky Makes You More Creative 101

Where Does Knowledge Come From?..............................103

The Value of a Good Cask 105

Is the Glass Half Full or Half Empty? 109

Toasting and Charring 111

Variety is the Spice of Life 113

Aligning Your Energy With Your Itness...Light It Up 117

Whisky Evaporates Faster Once the Bottle is Open!.........119

Three Plus Six Equals Nine! 121

Access...123

Meaning of Life..125

The Secret of Whispering Whisky 127

Whispering Whisky is Beyond Fun! 129

Not on Mondays 131

Final Note 133

Recipes...135

Last Thoughts..143

Author's note...144

Acknowledgements

Dr Ted, Educational Chair of a UK Forestry Society, who was one of my muses for this book, and was responsible for the co-creation of my wandering whisky whisperer's persona. Thank you my friend.

Dr Den, my research scientist friend, and fellow whisky tasting connoisseur, the smartest person I know. You have inspired me in every step of this writing journey. Your insights and editing patience were invaluable. This book would not have been written without you.

The lovely and extremely talented Cynthia, my award winning, best-selling author wife, whose support and guidance, I am truly grateful for. You made this book happen.

Neil, retired lecturer from a forestry school in Scotland, and whisky gentleman, who introduced me to some extremely special and very rare cask single malt expressions from Jura, Dalmore, Glenmorangie, and other unknown distilleries. Those whiskies were so glorious that they could easily have turned me from whisky whisperer into a whisky drinker, had they been more readily available. My time in Scotchland was made beyond epic by your generosity, your wisdom and your guidance.

My Irish friend, neighbour, and whiskey man Denis, who has expanded my knowledge of Irish whiskey and Irish culture. Cheers my friend!

To all the whisky master distillers out there, who have spent years and even generations perfecting their craft, I want you to know that your efforts and skills are truly appreciated. Many folk around the world are grateful for your talents. I am one of them.

To all my fellow whisky whisperers, who have graduated from my whispering whisky master classes, for your support and sharing of your musings, and yourselves, I am very grateful. Some of your ideas and quotes have been included into this book.

To Davin de Kergommeaux, author of "Canadian Whisky, the Portable Expert", a great whisky book, for your guidance along my Canadian Whisky journey. You are a true champion of Canada's finest spirits. I admire your passion.

And finally to Iain Banks, who wrote "Raw Spirit", In Search of the Perfect Dram, which has been and continues to be a source of my inspiration. I salute you at each of my tastings. May you rest in peace with the perfect dram by your side. I highly recommend his book to all of you who are on a similar pursuit in this lifetime.

Preamble

I need to mention, now, that I do not drink whisky. I merely taste whisky.

When you drink whisky, you get drunk and that is both stupid and disrespectful. When you taste whisky, it is much more egalitarian. It can open up a whole new appreciation. It is considered the proper way to experience this fine beverage.

However, when the mood is right, I like to go beyond merely tasting whisky and delve deeply into the act of whispering whisky, a concept which I developed, while living in Scotchland for 6 months and from spending many years in the world of whisky.

Whispering whisky is like an enlightened, mindful tasting experience, where you can develop a very deep and profound connection with the whisky you taste, the people you taste it with and more. When you whisper whisky, you develop respect and appreciation for many things.

- The skill of the master distiller.
- The skill of the cooper.
- The people you are at the tasting with.

- The time the whisky spent maturing in the cask.
- The connection the whisky has with the oak.
- Where the oak came from.
- What was in the cask before the whisky?
- The time the whisky spent getting ready to be tasted.
- How the whisky is feeling.
- What messages and lessons you get from the whisky.

This all builds a deep, true appreciation of the whisky and this then transfers to life in general.

Having said all that, I do swallow...the whisky I mean. If that makes me a whisky sybarite, so be it.

Introduction

Whisky teaches us lots of things.

Just like all good teachers who learn as much from their students and the content they teach, as their students are learning from them, you too, can learn many things from the folks at a whisky tasting and the whisky itself, if you are open.

So my friends, and whisky whisperers, be open, stay open; to the possibilities.

Be open to the many types of whisky and the numerous expressions from many countries.

Be mindful of the way that many expressions come into your life, as well as the people you share the whisky with and be grateful.

This is an awesome, magnificent and glorious journey.

Somewhere in all this, there is the perfect whisky, created just for you in that perfect moment! I hope you find it. I have found it several times but then things change and it is gone, in an ephemeral sense.

Humans are known as spiritual beings inhabiting human bodies and collecting thoughts, wisdoms, knowledge, and experiences.

Spirituality refers to oneness, connectedness, a journey in search of the sacred, or a blissful experience; all of which I have experienced while whispering whisky, which I am about to share with you.

Some say we are just two legged thought receptacles. These are some of my thoughts.

This book is a collection of my thoughts, spiritual guidance, shenanigans, and some ephemeral secrets, gleaned from my enchanted journey into, my interaction with, and connection to, whisky.

These are my wandering whisky whisperer musings, so to speak.

But first, I feel it is important to explain to you my whisky journey, provide you with a warning, some definitions, and the amazing complex maturation process of Scotch whisky.

I have also included how whisky has shaped my own maturation process, how to whisper whisky, along with the musings from various whisky whispering events.

At the end of the book, I have included a few of my favourite whisky based recipes.

I hope you have as much fun and enjoyment reading my book as I have researching and writing it! Enjoy your journey!

Cheers!

John Clement

My Whisky Journey

I began my journey to become the wandering whisky whisper back in 1977 when I was in first year Forestry at the University of Toronto. I was introduced to Chivas Regal at a party. I thought it was the coolest thing to drink and I still enjoy some of their more aged blends to this day.

However, none of my classmates liked it at the time, which I used to my advantage.

In 1979, I lived with 4 other guys in my class who all enjoyed Canadian rye whisky. I still enjoy Canadian rye whisky and am a fan of Crown Royal Northern Harvest Blend, Jim Murray's Whisky Bible world whisky of the year for 2016, as well as some of the higher end Forty Creek expressions.

Back then, though, if you had a bottle of rye in your room, it was fair game. When my room mates ran out and were still thirsty, they would help themselves to mine, especially if I wasn't home. But as none of them drank Scotch, none of my Chivas Regal got pinched.

In the late 80's my friend did his PhD at a Scottish University. He brought me home a bottle of Lagavulin 16 year old, which to this day remains my favourite smoky peaty expression, an Islay classic, followed very closely by Ardbeg 10 year old.

In the mid to late 90's, another friend had a New Year's party. That year, the Liquor Control Board of Ontario introduced the six classic single malts in full bottle sizes (750 ml) and the 50 ml airline bottle sizes. I got a full size Lagavulin 16 year old and he got a full size Dalwhinnie 15 year old and we each split the rest in airline bottles. We had a glorious time tasting the different expressions and writing tasting notes.

My taste buds at the time were not as developed as they are now, but everyone has to start somewhere.

Our tasting notes from that event were simply hilarious. Somehow, brown became a smell and "leave me alone you f***ing bastard" ended up in my notes.

My favourite two whiskies from that night were, you guessed it, Lagavulin 16 year old and Dalwhinnie 15 year old. (Although the other ones were all so glorious.)

The next 10 years saw me hosting and attending numerous tastings with many fine folk, where I continually worked on fine tuning my nose and palate, recording tasting notes on a diverse array of Scotch whiskies.

I also went to many wine and beer tastings and have toured many wineries and breweries.

I love to visit new places and have new experiences and learn interesting things along the way.

In 2014, a new neighbour from Ireland introduced me to the joys of Irish whiskey.

I introduced him to the joys of Scotch whisky. It is nice to have a wonderful neighbour who shares a passionate appreciation for whisky.

In 2015, I was able to realize my dream of living and working in Scotland. I spent a total of almost six months there, with the exception of 2 weeks in Ireland, a week in Cumbria and 6 days in Wales.

I sampled whisky where ever I went, and whenever I could, always writing tasting notes. (Perhaps another book, who knows?)

The first 6 weeks of this adventure, from mid-July to the end of August, was with my family, on an epic UK vacation.

The rest of the time in Scotland was spent working as a visiting lecturer at a Scottish forestry school, while on a "Sabbatical" from my home forestry college in Canada. A Sabbatical is a post-secondary educational term, derived from Christianity, which means to rest in the 7th year.

In biblical times, this would have been known as spending time in the wilderness.

Yet, I would hardly call anywhere in Scotland wilderness, although there are some pretty wild places, both literally and figuratively and yes, I did have some pretty wild times in some amazingly wild places with some amazingly wild folk.

When I wasn't teaching, or prepping or marking, I was researching whisky.

In my spare time in Scotland, I managed to taste over 70 whisky expressions, and visit 17 distilleries in search of the perfect dram.

So much for the rest part of the sabbatical. I worked and researched like a dog.

To say that I am fascinated with whisky and the interaction of spirit and wood is an understatement.

My passion for whisky and wood and forestry and learning about the influence of wood on whisky is beyond legendary.

Being a whisky connoisseur, an advanced level Toastmaster, an avid traveller, an award winning forestry educator, and a lifelong learner has provided me with the skills to delve deeply into, and share with you, my musings derived from my fascination of the interaction of spirit and wood.

Spirit, in this case, refers to the clear raw distilled alcoholic liquid that eventually derives its colour, smell and flavour from the oak maturation process. (More on this later.)

I also made incredible contacts with many whisky folks and even some whisky nobility, all who generously shared some of the most glorious Scotch whisky on the planet; some special cask bottlings that are not available at any price and some that are available for an astronomical price.

I once nosed and wrung 3 drops of a Scotch whisky from a bottle, from a previous client of a fishing guide, that sold for UK£75,000.00.

To all you whisky folks, on both sides of the pond, who so have been so generous to me in my whisky whispering journey, who have shared your whisky, your homes, your food, your stories, your passions, your ethics, your culture, your friendship, and your values with me, I am beyond grateful.

The following is a collection of my thoughts, lessons learned and musings that came to me before during and after my wanderings. I will continue to collect these insights, as they unfold, in the quest for the perfect dram, as I am the consummate lifelong learner.

To my critics, yes, I have probably killed too many brain cells in this pursuit, but if that is the cost of doing this kind of research, I'm okay with that.

I am in a constant search for the perfect whisky. I have launched a Facebook site called the wandering whisky whisperer. (And am soon to launch a blog site, a web site and perhaps make some videos.)

I also conduct whisky whispering master classes and tasting parties with food pairings.

If you know of a perfect whisky, please let me know. If you can provide me with a wee sample, I will shamelessly accept your generosity.

Warning!

This book will change the way you experience whisky.

Many of the musing contained herein were developed under the influence of fine whisky and may not make sense to those of sober mind. Also if you do not appreciate whisky, this book is not for you.

The ideas in this book have come from my musings and wanderings in pursuit of the perfect whisky, mainly while my friends and I, while not totally under the influence, may have progressed a wee bit down that path, which tends to be a natural but unintentional culmination of tasting and whispering whisky.

Whispering whisky is also fun!

Therefore, this book is meant to be read while slightly under the influence of a fine whisky or two, the finer the better. So before you continue, go and pour yourself your favourite dram and enjoy.

Always remember to respect the whisky and never drink and drive.

Also if you have noticed, I am using the term whisky, not whiskey. This is because I live in Canada and I studied Scotch whisky in Scotland and all over the world, for that matter. I am still on this journey.

I read somewhere that if there is an "e" vowel in the country you live in, that the word whisky is spelled with an "e". Not sure if this holds true everywhere but those whisky drinkers in the United States spell their whisky as whiskey and so do the Irish from Ireland. The folk from Scotland and Canada spell their whisky with no "e".

Interesting.

Finally, this book is a collection of insights and experiences along my life's journey. They are presented here to entertain and enlighten and educate those on a similar whisky journey in an "edutaining" fashion.

"Edu-tainment" is a relatively new term derived from a combination of education and entertainment.

I am a lifelong learner and edutainer and believe the facts included in these musings to be somewhat true, somewhat profound, somewhat whimsical, somewhat inspirational, somewhat entertaining, somewhat educational and somewhat based on my experience to date. (Heavy emphasis on somewhat.)

Having said that, my ego is not bothered, too significantly, if you, the reader, who holds other opinions or other pieces of life's puzzle, that I am currently missing, or who are further down the whisky path than I, would be so kind as to enlighten me.

I just ask that you do so in a gentle, loving and respectful manner. Always be kind, gentle and respectful, even when you are under the influence or hungover. Whisky has taught me that, several times.

What is Scotch Whisky?

First a definition...

To be officially called or defined as "Scotch Whisky", the raw spirit must be made in Scotland from 3 ingredients, malted barley, yeast and water, and be stored (matured) in oak casks in Scotland for a minimum of 3 years and one day.

Some whiskies can sit in the casks for up to 50 or more years for the maturation process to be complete.

Once it is bottled, it no longer matures. So if you purchased a 12 year old whisky in 1995, and open it in 2015, you are not drinking 32 year old Scotch, you are still drinking a 12 year old whisky.

Another Definition...

"Angel's Share" is the bit of whisky that evaporates yearly...approx. 2% per year thus explaining why as whisky gets older, it costs more. Also makes for happy Angels.

Question? If this is true, does a 50 year old Scotch have no whisky left in the barrel?

"Angel's Tears"- the part that runs down the glass, also referred to as "legs". This is the part that the angels didn't get and it looks like tears running down the inside of the glass.

A dram, by definition, is a moderate pour of a whisky that is pleasing to both the host and the guest.

Whisky Maturation Process

It follows then that maturation process is what is happening when the raw spirit, produced from the distillation of the three ingredients, malted barley, yeast and water, essentially distilled beer, is stored in wooden oak casks.

This is a glorious and complex interaction of the raw spirit and the wood that I, as someone with a degree in forestry, and whisky connoisseur, am most fascinated and intrigued by.

The complexity is overwhelming. I will keep this simple as the organic chemistry is difficult to comprehend and I know just enough to really appreciate the complexity.

There are 100's of possible flavour combinations in whisky which come from the ingredients, the processes and the interactions of the raw spirit with the oak.

Some of the nastier flavours are removed through the filtering effects of the charred layers inside the oak barrels.

Other flavours in the whisky come from the interaction with the oak wood in the casks and some flavours are imparted to the whisky from what was previously in the casks before the whisky was added.

These are all happening at different rates due to casks imparting flavours at different rates and also where the casks are stored in the warehouse and where the warehouse is located, and where the trees grew that make up the casks and how fast or slow the trees grew and which species of oak they were from and how the casks were prepared and on and on.

Storage temperature also influences these processes.

When the whisky and the oak in the barrels heat up, the casks absorb more whisky. When the barrels cool, the whisky leaves the wood and joins the whisky back in the casks. This in and out process adds flavours from the casks to the whisky and filters out impurities and creates chemical reactions with other chemical components in the whisky to produce the new flavours which ultimately build the unique character and brand expression of each whisky.

The complexity here boggles the mind, as there are way too many interactions and processes going on which I will lovingly refer to as "alchemy in a glass". It is this "complex alchemy" that I am truly grateful for and provides me with endless fascination and research.

Let's look a little closer at the oak.

Basically three kinds of oak are used in whisky making; American white oak (*Quercus alba*) and European oak (*Quercus robur*) and Spanish oak (*Quercus petraea*).

If this maturation process wasn't already complicated enough, let me add some more complexity.

Each species of oak has a unique chemical makeup.

Each tree within each species of oak, also has unique chemical components, picked up from the site it is growing on.

Add in that the staves are cut from different parts of the trunk of the tree, and can be from any of the three species in the same cask, in different proportions, adds yet another dimension of complexity.

How quickly or slowly the trees grow increases or decreases maturation time and if that isn't enough, how the barrels are treated, i.e. charred or toasted, will influence the flavours.

All these factors influence the whisky in a unique way.

Mind boggling or what?

Each distillery's raw spirit, is made from a similar but unique process, adds more complexity as does what was stored in the barrels before, (Bourbon, Sherry, Rum, Port, Red Wine, etc.), and if the casks were used by the distillery before...i.e. first fill, second fill or third fill casks.

Most casks are usually not filled more than three times.

These combinations in the casks are carefully monitored to determine when the whisky has reached its peak maturity.

The main storage of whisky is in ex bourbon barrels made of white oak from the United States.

However, some of the whisky will be put into ex sherry casks made from Spanish oak to finish the whisky, allowing the master distiller to get the distinctive character of their brand that they are after.

More complexity.

Some of the distilleries are starting to use virgin wood barrels.

These casks are brand new and have not had bourbon or sherry in them previously. Benromach Organic uses such casks, and this allows for a certain purity of wood flavour.

My concern was that the rate of maturation would be too quick for the whisky to be any good but the tour guide at Benromach assured me that the master distiller has taken that into consideration and after tasting the Organic, I would have to concur.

This is why many brands have gone to a no age statement and although I am not a fan of this trend, Scotch whiskies in virgin barrels can be glorious at as early as 6 years of maturation.

Oak is considered a ring porous wood.

In white oak, the pores are clogged with something called, "tylosis" which resembles fibrous spider webs.

Spanish oaks have them as well, in lower proportions, but the pores tend to be proportionally smaller to begin with.

Red oak, *Quercus rubra* and some of the other oaks that do not have tylosis, are too porous, and would have much more Angel share and would not be suitable for whisky maturation.

The species of oak used, not only impacts on smell and flavour, but also on colour to the whisky.

Ooh, the complexity.

Simply mind boggling, hence, the source of my endless fascination with whisky.

Then, the whisky, in constant contact with the oak in the inside of the barrel, sits quietly for some 3 to 20 plus years, maturing and waiting until the master distiller, with the nose of a blood hound, deems it ready and worthy of sale.

Some say the whisky goes to sleep.

Others say, "It is like a caterpillar that morphs into a beautiful butterfly".

Perhaps a delicious butterfly is more appropriate but whatever your analogy for the maturation process is, this is what makes whisky so variable and the perfect whisky so elusive and magical, just like the quest for the Holy Grail.

My Own Maturation Process

My own maturation process has been a bit less quiet, but none the less complex.

My mother died when I was 6 years old and my father died when I was 23.

I have had the help of many good people in my life.

Over time, they have protected me, mentored me, tormented me, guided me and enhanced the unique look, feel, smell and flavour of my personality and being, which is similar to how and what oak does for whisky over time.

Just like my friends and family with me, the oak brings out the finest a whisky has to offer.

As humans interact with the whisky spirit, their maturation processes intertwine and whisky attempts to influence that maturation process.

Whisky can increase your maturation process, if you let it. If you are not careful, though, whisky can also reverse your maturation process. Be wise to this.

Also the interconnection of spirits can make both spirits begins to soar.

Whisky makes my spirit soar like an eagle. Have you ever soared like an eagle? I usually do every time I whisper whisky.

How to Whisper Whisky

The term whisperer is, by definition, one who connects with something on a profound and deep level.

Like the dog whisperer, the horse whisperer, or the ghost whisperer, the whisky whisperer must make a profound and deep connection with the whisky, in order to understand how it feels. Only then can you hear what the whisky has to say.

This connection is then shared with those in the room, at the tasting, which leads to stimulating conversation and tremendous insights into life, which make the whisky whisperers feel enlightened, and this makes the whisky happy. It also deepens the connection between the tasters in the room.

To make this happen, I must clarify a most important point.

When whispering whisky, you don't drink whisky, you taste it in a mindful way. And, it is important to swallow to get the full whispering experience.

I don't believe in spitting good whisky. That would be tantamount to alcohol abuse. And that my friends, is the story I'm sticking to.

In order to truly whisper whisky, you must experience the whisky with all your senses.

In the many tastings I have been to, and conducted, and in every whisky whispering master class I have taught, I always pick up little bits from the whisky, and others, along the way.

- Have a book to record your tasting notes and whisky inspired ideas in.
- Be open, stay open, to the ideas that come up and record them.
- Have fun. Whispering whisky is fun.
- Researching whisky is fun! Try as many expressions as you can.
- Read as many books on the subject as you can.
- Watch YouTube videos on tastings.
- Be humble and do not pretend to be the definitive expert on the proper way to taste whisky. In that, learn from others at the tasting.
- Enjoy the learning journey that you are on.
- Enjoy sharing your insights with others.
- Enjoy others' insights that they have gained through their process.

- Connect with others at the tasting, in a deep and meaningful way.

Whispering whisky involves tasting it with all of your senses.

Sight

The eyes begin the process, enjoying the various colours, and the formation of the "legs" on the inside of the glass.

Smell and Feel

The nose takes over, and as the whisky tickles the nose, it then prepares the mouth.

Taste and Feel.

The mouth does double duty here, both tasting and feeling the whisky. Many flavours come out. Then, there is the burn. Does it linger? Is it pleasant or unpleasant?

These sensations are all noted in your tasting note book.

Listening

But where does the listening come in to play?

I believe that once the other senses have been engaged in the process, the wisdom of the whisky begins to speak, and shares with you and the others at the whispering, untold insights into the nature of worldly affairs, but only if you are open to listening for them.

Hence the true whispering part. Then, the sharing of the insights creates even more insights and on and on it goes.

Enjoy this sharing. Enjoy your moments. Enjoy!

Let's Start

To get started on the whisky whispering experience, you first have to have the proper whisky glass.

The simplest way to describe the perfect glass for whispering whisky is that it must resemble an upside down light bulb in a sense.

You want the conical bulbous bit to hold in the volatiles, so that they are available for nosing. There are a wide variety of whisky glasses on the market but a straight sided glass or a martini style glass just won't cut it.

A brandy snifter or a stemless wine glass will also do in a pinch, although some tasters might prefer a stemmed wine glass.

A frozen beer stein or anything plastic would be very disrespectful of the whisky (and the folks that made it).

And, mark my words, you do not want to disrespect whisky, ever.

Before I begin to whisper whisky, I like to have a few sips from a proper oak aged Scottish beer to cleanse my palate from whatever I have eaten previously and to wake up the taste buds. It sets the mood and allows the connection to begin.

Now pour a wee bit (about 1.5 cl, (15 ml), or ½ oz.) of whisky into the proper glass.

At first, I thought it prudent to use a finely calibrated shot glass, but after a while, you get pretty good at pouring the proper amount into a glass. This will give you enough for 3 small sips.

Some folks say to put 4.5 cl, (45 ml) or 1 ½ oz. into a glass for a tasting but I find that is too much if you are going to do multiple tastings in an evening.

Then, swirl the spirit around in the glass. Warm it up with your hand. Appreciate the colour.

Appreciate that the colour of the whisky comes from the time it spends in the barrel in contact with the charred or toasted oak.

Some distillers add caramel colouring to the spirit but I personally don't like that. Let the whisky be natural. I also realize that sometimes colour is added to maintain a consistency between batches and I can at least accept that.

The darker amber colours tend to come from the sherry casks while the yellowy gold colours come from the white oak bourbon casks.

Also, admire the legs of the whisky. The longer and thinner the legs, the less viscous the whisky and some say the younger the whisky. I have also heard that the legs are known as the "Angel's Tears" because that is the bit of the whisky the Angels didn't get as part of their "Angel Share".

Swirl some more. By now, the vapours have woken up.

Time to wake up your nose.

Take a wee whiff at the top of the glass but don't go too deep too fast. You don't want to hurt yourself.

Nose in deeper as your nose gets acclimatized to the strength of the vapours. You have to be gentle on your nose, because you will want to save it for future tastings.

As the nose begins to connect with the vapours, note what smells are you receiving from the whisky?

Can you pick up caramel, butterscotch, toffee, vanilla, citrus, pepper, ginger, raisins, berries, strawberries, cereals, iodine, leather, oakiness, florals, smokiness, peaty smells, etc?

Now swirl a bit more and when the moment feels right, nose in even more. Slowly begin to inhale with your nose in the glass.

This begins an amazing journey with whisky.

In fact, some of my tasting notes say I could drink this whisky with my nose. The Benromach Organic and the Glenfarclas 15 year old and the Crown Royal Northern Harvest Blend are three of my favourite whiskies to nose but if you every get up to Highland Park in the Orkney Islands north of Scotland, do the tour. I hope they still allow you to nose the cask from 1968.

That was one of the highlights of my wanderings.

After the nose is engaged, you are almost ready to engage the taste buds but not quite yet.

Now, nose the whisky with your mouth open. You heard me right. This is not easy for some as they cannot rub their bellies and pat their heads at the same time.

But do this and you will be pleasantly surprised.

The sides of your tongue are starting to wake up and even though you are still nosing the whisky, you will begin to taste it. After all, the nose is connected to the taste buds.

I guess that's why if you don't like the taste of something, plug your nose and taste it. But please don't do this with whisky.

I did learn this last bit from my friend Sam at the Whisky Castle in Tomintoul. If you ever want to taste whisky before you purchase it, visit the whisky castle. This is a must stop at place if you are ever in Scotland. I found this place several times on my various whisky wanderings.

Make sure you have a designated driver, as there is very low tolerance for drinking and driving in Scotland. In fact, always have a designated driver for every tasting you do, anywhere.

Once the taste buds are woken up, it is almost time to taste.

Enough whisky foreplay already!

My method is to swirl once more, appreciate the colour with your eyes, quick smell, appreciate the smell and tilt your glass and let a wee bit enter your mouth at the front of the tongue, chew it for about 3-4 seconds. I am not a fan of leaving it in your mouth for 1 second for every year old like some folks do. That would mean chewing or swishing it for 20 seconds for a 20 year old and that is just too long. 3 - 4 seconds in total is fine, and let it slide down your throat slowly and find its own way home.

Start to appreciate the mouth feel.

What did the whisky just do? Did it come in sweet, dry, is it hot? Does it burn? Does the heat linger? Was it smooth or harsh? Did it go down, come up, go down come back up and smack you upside the head? Or did it behave and slide down, warming you up and then disappear.

Do not gulp, for if you do, you will feel the burn on your throat.

And you don't want that. Slowly let it go down and find its own way home, so to speak. After all, some of the whisky you taste has taken years to mature and you don't want to rush this.

The first sip is mainly to wake up your taste buds and prepare for the next sip.

Now that your mouth, your tongue, and your taste buds are fully awake, it's time for the second sip.

Aww, the glorious second sip.

This is where you realize the full glory and taste of the whisky.

My favourites are the whiskies that the second sip behaves completely different in the mouth than the first sip, but both sips are pleasantly playfully dancing in your mouth, and continue giving of their secret, hidden flavours.

The game now becomes...guess what flavours you are tasting.

Other Scotches behave the same between sips, and the second sip actually gives you exactly the same expressions of the flavours and the mouth feel as the first sip. These Scotches in my opinion are boring and predictable.

Some whiskies, the first sip is the finest, the second sip is worse and the third sip is even worse. These are my least favourites.

Now for the third and last sip, add one drop of water. You heard me! I know there are folk that think this wrong but there are folk that think this right, and I am one of them.

Just do it!

You may be pleasantly surprised by the impact that one drop of water has on the whisky. Or you may be disappointed. But it will definitely elicit a reaction.

Swirl it around and let it stand. Some say to let it stand for 15 minutes to let the water fully integrate into the whisky at the molecular level.

Even though I am a patient man, this is way too long. One – two minutes is about right for me. 15 seconds at an absolute minimum.

While I wait for the water to do its alchemy, I use that time to connect with my fellow tasters or begin writing my notes. (More on that coming up.)

I know, some of you purists are cringing, you must wait the full 15 minutes, just as some of you are cringing at the thought of adding water.

I have found the stronger the alcohol by volume ABV, in the whisky, the more water you can add until it is just right for you. I still add one drop of water at a time and use a fancy glass pipette I got at one of the distillery visitor center gift shops. A simple teaspoon also works just fine.

Some say to add enough water to get to 35% ABV for the best taste, others say 27% ABV dilution is the optimum. Too much math for my liking. I prefer to add a single drop of water when I am ready.

I must add at this point that none of this nonsense is mandatory.

Any way you taste whisky is fine, as long as you don't put it into the freezer, pour yourself a frozen shot and slam it like you would a frozen vodka. That would be bad, very bad.

In fact, I dare say that those who have done that probably don't like Scotch or any whisky at all.

Heck, they may not even like themselves. It would also be disrespectful of the whisky and you never want to disrespect whisky.

Once you have done all these steps, you are ready to whisper the whisky.

It is time to be mindful. It is time to be open. It is time for connection.

Be in the now with your whisky and your friends. Truths will appear to you, in your thoughts, or from the thoughts of others at the tasting, as a result of all this preparation.

Remember it is the whisky that is whispering these truths to you, sometimes directly or sometimes indirectly, through your friends at the tasting. Be open, stay open.

The above mentioned procedure is my way of whispering whisky, which allows for a deeper, more mindful respect of both friends and whisky and life in general.

Be sure to write notes as you go. Record the date, time, place, and who you were with. Make sure you get a book that you will bring with you to all your tastings.

Start your notes when you view the colour. Next is the nosing.

In one of my more vigorous tastings, when I reviewed my notes, the next day, apparently, as I mentioned earlier, brown is a smell; made sense at the time, hence the warning at the beginning of the book.

Record the flavours you taste and the truths about life that may emerge, either from your interaction with the whisky or your interaction with each other.

This is the way I taste and whisper whisky. Enjoy! Cheers!

The Musings of the Wandering Whisky Whisperer

The following are insights, known as musings, which have come to me while whispering whisky. I hope you enjoy reading them as much as I have enjoyed writing them.

John Clement

Respect

Whisky teaches us respect. You must respect whisky or you will be in big trouble.

It is that simple.

This is the most important lesson to learn from whisky.

I don't drink whisky anymore. (I have in the past.)

I now think it is disrespectful to drink whisky.

I taste whisky and love the challenge of trying to figure out what the flavours are and where they come from.

I do not spit whisky and cannot figure out why anyone would.

I do research. I read reviews. I watch you tube videos.

I visit distillery web sites and when I can, the actual distilleries themselves.

I mainly taste whisky and write notes. I have a different style of tasting notes and may publish them one day.

On the rare occasion in the past, when I have drank whisky, I have truly learned that if you forget to respect her, you do pay a heavy toll for that momentary lapse.

We all have learned this lesson the hard way and it seems that some of us have to keep learning this lesson.

Smarten up.

This happened to me a few years ago.

I had some friends over for a celebration. As a special treat to mark the special event, I purchased a glorious 18 year old whisky and we did a bad thing...(which seemed like a good thing at the time) which was to open the bottle and throw away the cork, a practice I did learn from a real Scottish lad, who by the way is now deceased. Hmmm.

The result was horrible.

For me, it was an awful hangover.

But at the time, I had had a cold and a nasty cough it was. I just couldn't shake. Two doctor visits and 2 courses of antibiotics later and still this cough persisted.

However, that one fitful hungover morning, after a night of flagrant disrespect of whisky, and my cold was gone.

Thank you whisky. Not sure the cure was worth it though.

Some of my mates didn't fare so well.

This is what can happen when you do not respect whisky. I am sure all of you have had a story or two about when you disrespected the whisky.

Some have lost everything because of this.

Don't let that be you.

I have found that in my tastings, if I have 3-4 whiskies, the tasting is civilized. That allows for a couple of extra whiskies that call out to you to be sampled.

I do believe that the Scotches you have at the back of the cabinet need to be opened and tried but save some of them for special occasions. I have a couple of whiskies, at the back of my cabinet that will be opened one day for special occasions.

My father had a saying which he repeated to me on more than one occasion and that was, "*If you want to play, you have to pay*". He never told me I'd have to pay so much or pay so often.

I do really like whisky for that lesson that it imparts to you every time you taste it.

If you respect the whisky, it will take you places you cannot even imagine.

They say, "IN VINO VERITAS," which translates to "In wine, truth." This loosely means that when you drink, you tend to speak the truth more. It is amazing what truths are revealed when you taste and appreciate whisky with good friends.

You just cannot script this stuff...

The inspiration and musings for this book have come directly from a gently, and sometimes not so gently, altered reality that has come from respecting the whisky.

If I hadn't respected the whisky, none of this would have been possible.

So my advice is to always respect the whisky!

Always!

Respect!

On the topic of the respecting of whisky, Judge Noah Soggy Sweat said it best back in 1952, and this has been the basis of political double speak ever since. It was at a Toast Masters' meeting that I'd originally heard it. It captures, in a brilliant fashion, the pros and cons of whisky.

Here is the famous, if not infamous speech, "If by Whiskey Speech" by Noah "Soggy" Sweat:

"My friends, I had not intended to discuss this controversial subject at this particular time. However, I want you to know that I do not shun controversy. On the contrary, I will take a stand on any issue at any time, regardless of how fraught with controversy it might be. You have asked me how I feel about whiskey. All right, this is how I feel about whiskey:

If when you say whiskey you mean the devil's brew, the poison scourge, the bloody monster, that defiles innocence, dethrones reason, destroys the home, creates misery and poverty, yea, literally takes the bread from the mouths of little children; if you mean the evil drink that topples the Christian man and woman from the pinnacle of righteous, gracious living into the bottomless pit of degradation, and despair, and shame and helplessness, and hopelessness, then certainly I am against it.

But, if when you say whiskey you mean the oil of conversation, the philosophic wine, the ale that is consumed when good fellows get together, that puts a song in their hearts and laughter on their lips, and the warm glow of contentment in their eyes; if you mean Christmas cheer; if you mean the stimulating drink that puts the spring in the old gentleman's step on a frosty, crispy morning; if you mean the drink which enables a man to magnify his joy, and his happiness, and to forget, if only for a little while, life's great tragedies, and heartaches, and sorrows; if you mean that drink, the sale of which pours into our treasuries untold millions of dollars, which are used to provide tender care for our little crippled children, our blind, our deaf, our dumb, our pitiful aged and infirm; to build highways and hospitals and schools, then certainly I am for it.

This is my stand. I will not retreat from it. I will not compromise."

I love this speech for many reasons. Being an advocate of respecting whisky, I truly appreciate how Noah Sweat cleverly illustrated the pros and cons of whisky.

Being advance level toast master, I can appreciate the humour, the cleverness of the flow and the duality of this speech.

There is duality in lots of things in life, including whisky. Although, life and whisky are far too complex to limit them to a simple duality.

Just remember to always respect whisky!

John Clement

Patience

Patience is a very useful virtue to possess and work on and develop.

My friends have told me I have the patience of a saint. (My critics will tell you I have the patience of a Scotch Whisky Distillery Visitor Guide in August.) Either way, as a result of my whisky whispering wanderings, I have learned and developed greater patience by reflecting and musing about the maturation process.

Whisky must sit quietly for at least 3 years and one day before it is considered whisky, at least Scotch whisky and this maturation process must occur on Scottish soil, in oak. Some whiskies have to take up to 20 or more years to reach their true potential. (See the maturation process above)

If you were to sit quietly for that long, it would require a great deal of patience. It would define patience.

Patience is a virtue and being a reflective learner, I have learned many things by patiently reflecting on them. (Just like the whisky does during its maturation process.)

Whispering whisky requires patience to allow the fullness of the sensory experience to properly unfold.

Patience also allows you to stay open to the possibilities and opportunities that come your way.

Taste whisky. Whisper whisky. Be patient. Have respect! Cheers!

Taste the Difference

There are so many different whiskies out there that I am convinced there is something for everyone for every moment. Having said that, one of my friends has tried to like whisky, on more than one occasion, but she just can't. She hasn't found that elusive one yet. I think those who haven't found the perfect whisky, have simply not put in enough effort or have just given up on the search.

I remember when I first tried Scotch whisky, back in university in the late 1970's, it was an inexpensive blend. Although it was quite an affront to my inexperienced taste buds, I thought that was the most glorious whisky in the world.

My taste buds have changed and evolved a lot since then and they continue to evolve and get more refined.

I would say that people will change in their appreciation of different whiskies as their tastes evolve. A cask strength whisky may be too harsh for a novice taster, but I am enjoying those more and more. Cask strength, non-chill filtered, yumminess at up to 61% ABV, and beyond.

Different whiskies also taste different to different people.

If fact, the same whiskies can taste different to different people. We are genetically different with respect to taste bud arrangements, and activation. Throw in the fact that our taste buds change over time and you can begin to appreciate the difference.

Let's look closer at the genetic diversity of our taste buds.

Take the Brussel sprout argument...some love them, and some hate them. A person who loves Brussel sprouts cannot understand why a person would not like them. I personally think they are vile!

The asparagus smell in urine is another interesting genetic diversity in the population. Some people can smell asparagus in their pee shortly after eating asparagus and some cannot.

This partially explains the diversity of preferences for different Scotches for everyone and it now begins to make sense why there are so many different expressions out there.

Take 10 year old Laphroaig, for example, a heavy, smoky, peaty whisky that my taste buds rebel against, every time I taste it. The ashiness is way too intense and lingering for my palate at this point. Yet I know many whisky folk who enjoy it, telling me it is their favourite. They enjoy the complexity and the lingering effects that they find truly magnificent, which might explain why it sells so well.

I prefer Lagavulin 16 year old if I am in the mood for a smoky peaty whisky, although recently Ardbeg 10 year old is making its presence known to me.

This difference can only be explained by a difference in individual taste buds.

There are apparently over 10,000 expressions of whisky, worldwide.

"Vive la difference" the French say. Enjoy!

Moments

What may be the perfect dram in this moment, may not be the perfect dram in the next moment. Many things influence our moments. What we eat? How much sleep we had? Are we healthy? Did we exercise recently? Is the water we drink pure? Is the air we breathe clean? Do we smoke, or vape? How healthy are our relationships? What is the weather like? What is our house like? Even who you are sharing a dram with, can influence your moment.

All this, and more, seem to impact on what we perceive to be the perfect dram at the time. This is why my top 10 Scotch list currently has 31 Scotches on it. It could even have more as I mentioned earlier, there are over 10,000 different expressions of whisky and I have at the point of writing only tried about 200. I am sure some of the ones I have not tried will make my top ten list, which will continue to grow and change. Perhaps this section could have been titled, "My top ten list has 31 Scotches on it."

I have also done tastings where I sampled 4 Scotches, picked a favourite, tasted them again, and came up with a different favourite, all in that same evening. Tasting whisky is just that complicated.

So my advice is to appreciate the complex influences of the moment. Be sure to include your friends and recognize their contribution to the moment.

Always enjoy your moments.

There are no perfect people, only perfect moments. But there is a perfect whisky for your perfect moment if you are open and mindful.

Allow whisky to shape your moments. Fine whisky will produce fine moments for and with fine people.

Allow whisky to teach you about your moments for she is a fine teacher, as long as you respect her. Cheers!

View From the High Road

"Ye take the high road and I'll take the low road and I'll be to Scotland before ye"...as the song goes.

Tasting and whispering whisky allows us to take the high road. (And sometimes the low road). Oh the duality of whisky.

The care and craftsmanship that goes into the production of fine whisky, also goes into us when we taste the whisky. This becomes apparent as we whisper whisky.

Master distillers have the nose of a blood hound and whisky making in their genes...most are multi-generational or have been reincarnated from such and have worked their craft to a high level. Always appreciate the skill of the master distiller.

Whisky allows us to be better folk. It always calls for us to take the high road. Although, sometimes we don't listen and take the low road.

I recently saw a show about Scotch whisky where the young lad said Scottish whisky producers were put on the planet for one purpose and one purpose only and that purpose is to produce the most glorious whisky possible. A noble endeavour for sure.

I believe his words to be true! Whisky allows us to be better folk. And that, whisky whisperers, is the view from the high road from which I speak.

Always take the high road, which is a metaphor for always do the right thing, instead of doing things right.

Enjoy the view from the high road which always seems a wee bit more glorious when whisky is involved. Cheers!

Scotchland

I think they should change the name of Scotland to "Scotchland", the land of Scotch. This one came to me from my son when he told his friends that my dad is over in Scotchland at the moment. This Scotchland is an enchanted place, which I often whimsically reside in. It's like I am transported to Scotchland, in both my mind and my heart, when whispering whisky

It is a delightful place or space to be in.

Is Scotland not the land of Scotch Whisky? Maybe someone should start a petition to rename Scotland to Scotchland.

Scotchland indeed!

Down to Earth

I occasionally enjoy a smoky peaty Scotch and have found that when doing so, the conversation with friends seems to have a deep, down to earth feel about it. I wonder where that comes from.

Could it be that connecting with the earthy peatiness and primal smokiness of the whisky influences the conversation? Is it some kind of a bioregional thing? Do certain regions of Scotland have finer peat than other regions?

If you are going to include a smoky peaty dram in your whisperings, do so at the end of the night, or only taste smoky peaty whiskies that night. The hit that your taste buds take after a smoky peaty Scotch will leave them useless for further tasting functions just as they claim that once a smoker stops smoking they gain weight because the food tastes so much better.

Therefore, leave the smoky peaty whiskies till the end of a tasting.

Also if you are a smoker, don't go out for a smoke during a tasting, if you can help it.

Enjoy the earthy peaty whiskies and see if this happens with you and your friends.

Grit

Whisky tasters seem to be the most successful people. Why is that?

Angela Lee Duckworth once said that the best characteristic that defines success is grit.

Tasting whisky properly is definitely a gritty experience and not for the faint of heart. This may be why whisky tasters are so successful. It takes grit to truly appreciate whisky. It is a necessary ingredient to those who desire to whisper whisky.

Who knew that another good predictor of success would be whispering whisky?

Take Note

I take notes when I do tastings.

I have a special orange moleskin book that my lovely and talented, award winning, best-selling writer wife and best friend, Cynthia A Clement, author, picked up on her first trip to Scotland at the Glenmorangie distillery along with a lovely bottle of Glenmorangie Astar.

As stated in the "How to Whisper Whisky" section, I like to record the date of the tasting, who was there, the name of the expression, the ABV (alcohol by volume), the colour, the nose (smell), the tastes, first sip, second sip and third sip with one drop of water added, the finish and any crazy thoughts that just fly out of my head or the mouths of those with me at the tastings.

That is where the quote at the front of the book came from. One of my friends named Juhani said "If you love whisky, whisky will love you" and at the time, I thought it was the most beautiful quote ever spoken.

Why does my hand writing get so sloppy and illegible towards the end of a tasting night???

It doesn't matter. Get a tasting book and use it to help you take note of your experiences and thoughts. Then you can review them and reflect on them, and share them if you like.

There are way too many expressions to keep track of. And how will you know which your favourites are? It is also a very civilized thing to do.

Maybe this is just the scientist in me. So my whisky whisperers, take note!

Appreciate the Finer Things in Life

Too many people don't appreciate whisky. They just toss it back or mix it with pop. They get drunk and disrespectful.

Whisky is meant to be appreciated. Heck, I will go so far as to say that whisky was made to be appreciated. Whisky gets sad when it is not appreciated.

Fine Scotch whisky is one of those things that teaches us to appreciate the finer things in life.

When someone first tastes any whisky, it is an affront to their senses, just like when you tasted any alcoholic beverages for the first time.

However, as one perseveres on this journey, they begin to pick out the complexities and after a while, they can begin to truly appreciate those complexities and the effort and the alchemy that goes into producing fine whisky.

As a result, they can begin to develop an appreciation for the finer things in life, which manifests itself into other areas of their life as it is like a transferable skill.

I once bought a 100% cashmere scarf back in 1982. At the time it was a $100 scarf on sale for $50. I have worn that scarf every year since then and I never tire of the feeling I get when I am wearing it.

One of the best ways to learn to appreciate whisky is to take a tour of a distillery. Don't do the cheap tour, though. Do the highest tour you can afford. The higher end tours are for the most part the ones with the best value for the money.

The Highland Park tour at their Orkney Island distillery was exceptional. My new best friend, Lee, the tour guide, on the Connoisseurs Tour, told us that we would be tasting a 12, 15, 18, 21 and 25 year old.

At the start of the tour, he proceeds to tell us that he is embarrassed that there is no 25 year old at the distillery but he will substitute a Scotch of equal or greater value and he says that with a twinkle in his eye.

While on the tour he tells us how wonderful the 25 year old is; the only Scotch to score 100 / 100 points in a competition. Then, he proceeds to shows us a cask with 1968 printed on the front and he pulls the bung out of the barrel and allows us to nose it. One of the guys on the tour and I lose all resemblance of civility at that point.

It was that amazing.

We try in vain to figure out how we can take that empty cask home with us, but both agree it might be too difficult to get that on the plane.

I could have nosed that cask until I die, it was that good.

Lee, the tour guide, then tells us that it was bottled in 2008 making the expression a 40 year old Scotch. He then tells us that would be the replacement for the 25 year old. OMG!!!

The 25 year old sells for about $900.00 (Canadian) at the moment, if you can get a bottle. In the distillery gift shop, the 1968 expression sells for about $6500.00 (Canadian), making the whole tasting experience at the distillery, an exceptional value. And a most gloriously fine dram it was!

Later that night, my friend and I were able to get a dram of the 25 year old at the Stromness Hotel. It was even better than the 40 year old. That night, the 25 year old expression became my number one all-time favourite whisky.

I cannot wait to try it again.

In fact, as mentioned earlier, I have 31 whiskies currently on my top 10 list, but right now, it is truly my number one. I just cannot afford it at this point in time because I have spent a wee bit too much money doing my research for this book.

I am planning to be able to afford it one day, so tell your friends to buy my book!

As mentioned previously, I have also tasted 3 drops of a $160,000.00 (Canadian) bottle of Scotch from a client of a fishing guide in Inverness.

I have tasted the once in a life time bottlings from special casks with those who are connected in the whisky industry. Some were better than others. Some were incredible.

For some, the story of how they were acquired was better than the actual taste.

I think that is the way with most of the finer things in life. They have to be earned to be truly appreciated.

However, I do know that there is a Scotch for everyone's taste and everyone's budget and a Scotch that everyone enjoys but cannot yet afford!

Always appreciate the finer things in life!

They make life worth living. So does whisky!

I recently read an article of a gentleman who had a shot of whisky every morning in his coffee. He lived to be 107 years old.

Whisky makes life worth living.

Whispering Whisky as Mindfulness in Tastings

Mindfulness is an expression that is gaining popularity and it is used in relation to some type of meditative connection.

Mindful meditation is to be fully present in the now, to be open to the fullness of moment in order to experience and appreciate life on a fuller basis.

One of my gifts is to see connections between things that others do not see. I don't know of anyone yet who has applied the concept of mindfulness to whisky tasting.

So let's take this concept of mindfulness and apply it to whisky tasting and come up with a new concept called "whispering whisky."

To me, Scotch whisky is the alchemy of the United Nations (UN) in a glass.

As you whisper whisky, you begin to appreciate this international alchemy and what influences the character (colour, smell and taste) of whisky; the softness and peatiness of the water, the type of peat, the type of malted barley, the malting process, the distillation process, even down to the dents in the still begin to shine through.

To truly begin to become mindful when whispering whisky, how all these combine is only the beginning.

There is also the wood which interacts and protects the whisky, allowing it to develop into the magic in a glass that it becomes.

Appreciate that whisky is made with the barley, the water, the yeast, the peat and it is stored in oak barrels from America, Spain, France (the UN bit).

Next be mindful as to if there was bourbon, or sherry in the barrels previously.

Was it a first fill or second fill cask?

Then think of the influence of the age of the cask, the age of the whisky, the rate of the cask to impart flavours and do its barrel stuff, the charring or toasting of the casks and the environment the casks are stored in... by sea air, up in the highlands or what part of the shelf the casks are stored in etc., etc., etc...

Once you begin to truly understand the complexity of each of these factors, you can begin to truly appreciate the whisky, which then allows you to begin to whisper whisky.

Whispering whisky is a mindful pursuit.

It grows like the barley, and the oak trees that are all part of the process, and this becomes part of your process.

Mindfulness in Scotch tasting and whispering sets the stage for the whisky to allow the alchemy to unfold.

Chemistry explains the catalyst for reactions to unfold. Whisky is the catalyst.

Alchemy explains the rest. Alchemy is also the catalyst to deepen the connection with your friends and to appreciate the finer things in life.

Be mindful that the peat of Scotchland is centuries old, the oak the casks are made from can be centuries old, and the maturation process of whisky with oak is centuries old.

The craft and the wisdom of the master distiller is also centuries old, handed down through the generations.

All this accumulated "old", materials, traditions, experience and wisdom, somehow impacts on us in ways we cannot imagine.

There are layers of meaning that come from whisky.

There are layers of complexity in a glass of whisky, that when the time is right, add to the complexity of the interactions of the humans at the tasting. "Whisky provides a pathway to the deeper meanings in life." (Matt, 2016.) (At one of my whisperings.)

The forest, like whisky, has a "uniquity" about it, antique uniqueness that has the answers to many of life's questions and this is where the whispering bit comes in. (I made up the word "uniquity" but it seems to be appropriate.)

Just be open enough to get those answers.

You also have to be still and relaxed and mindful to really get into the whispering.

There is infinite wisdom that comes from the forest and is transferred through the barrels into the whisky and into us when we taste it. Again, be open, stay open to the wisdom that may appear as a result of whispering whisky as this is what whispering whisky is truly about. (Along with an incredible amount of fun, and a slight bit of intoxication.)

This book has written itself after numerous tastings, it has sort of "whispered" itself to me so to speak.

Whisky influences my friends, my family, my critics, my enemies, and they all combine to make me who I am...and them who they are...and so on and so on...

Statistically on a normalized curve, you will have the outliers on both sides of the norm; both positive and negative types. Those outliers on the upper positive end are special spirits. Like great and glorious whiskies, these are the teachers that go beyond the normal to enlighten us.

They stand above the mass market, and are my favourites to whisper. They teach me the most.

Be a positive outlier!

Whisky makes us further aware of our spiritual side and connects us with a deeper appreciation of everything in life.

The oaks, the bourbons, the sherries, the water, the raw spirit, so many spirits come together in a glass... perhaps this is why whisky is known as a spirit.

Where does the heat of the whisky come from? Some would say the alcohol content...however some of the better whiskies with higher alcohol content burn less that the rougher ones with less ABV.

I believe the production of whisky chemically binds the heat from the sun as it runs through the barley as it is growing, and through the oak trees from southern US and Spain; this imparts the heat in the glass.

Be one with the whisky and it will take you to some magical places and allow you to experience those perfect moments mentioned earlier in the book.

I also find that once you go to these magical places, in perfect moments, you are also building memories that will last a life time. Perhaps this is what the true purpose of the whisky is while waiting all those years in a barrel.

I wonder how the whisky feels being left alone for all those years.

I also wonder what the whisky is thinking.

John Clement

Whisky Makes You Happy!

In his "If by Whisky" speech, presented earlier, Judge "Soggy" Sweat said, "whisky puts a glow of contentment in their eyes".

I have seen and experienced this "glow of contentment" on many occasions while whispering whisky.

I see it when folks get happy, right from the first nosing of a whisky at a whispering session. In fact, some whiskies smell better than they taste.

I could literally be happy just nosing some whiskies.

One of the things I like to do is to enlighten people and help them gain a brighter perspective on life. Whisky, if whispered properly, can certainly help with this process.

Whisky can also help you put a "hitch in your giddy up"!

After whispering a wee bit of whisky, some folk are known to be "lit up like a Christmas tree."

I always feel a lot happier when I whisper whisky and so do those who whisper whisky with me.

However, on occasion, I have met certain people that when they have a taste of whisky, the "ugly stick" comes out and they get mean and stupid. If you are one of them, do not do whisky, plain and simple.

Life is too short to be mean and stupid. Allow whisky to make you happy!

The True Value in a Bottle of Whisky

It is amazing what you can get for a bottle of whisky. I have known people that will trade just about anything for a bottle of whisky.

But what is the true value in a bottle of whisky?

I think it is priceless. The conversation that flows, the friendships that become closer, the good times that happen, strangers that become best friends, and the deepening of memories, and the understanding, all of which, add to the value of whisky.

The five billion dollars on average that are added to the Scottish economy annually and the taxation that funds many things in our community are also part of the value of whisky.

Whisky is used as currency in some areas, and as an investment in other areas. The problem for me with investment whisky is that I could never leave a bottle untasted in the back of my cabinet. Once it is open, it loses its investment potential. Darn!

The enjoyment you get out of a bottle of whisky is a bargain, even at twice its price.

Whisky is more valuable than you realize.

Whisky Will Protect You

The forest will protect you. Homes built of wood from the forest will protect you. Foods come from the forest and plants will nourish you. One half of all medicines come from plants and will protect you.

Whisky comes from plants and wood and process. Whisky will protect you. Whisky will cure you.

Doctors used to prescribe whisky to cure the common cold and I have found relief from pesky colds after drinking whisky.

I have also used whisky right after being bitten by a tick, as a topical disinfectant.

While in Scotland, I found a tick had burrowed into my arm. I used the olive oil and cotton swab spinning method to remove the tick. Google search for tick removal.

Then I took a small 5 cl (50 ml) empty whisky bottle and put about 5ml of the strongest cask strength whisky I had into the bottle, turned it upside down on the tick bite hole area, allowing the whisky to contact the bite site and left it there for about 15 minutes. This seems to work as a disinfectant and doesn't harm your skin.

I have also rubbed whisky on my balding head, but my hair didn't grow back.

Whisky is a good mouthwash though. Just don't drive right after you use whisky for mouthwash.

Whisky Makes You More Creative

Creativity that goes into the making of whisky transfers into your being when you drink it and ultimately makes you more creative, which is essentially how this book came into being.

Whisky is the creative juice that fuels the musing within this book and is responsible for this part of my creative journey.

Whisky provides the creative juice to whisper whisky.

Once you are in the creative flow and your creative juices are flowing, do not be afraid to be grateful, for this will enhance your abundance as well.

I wonder if this is why a lot of creative people have enjoyed whisky and fine spirits in the past.

Where Does Knowledge Come From?

Go into the forest and ask the forest for guidance. If you can be silent enough and can truly listen, you may be surprised at what might come to you.

Trees have lived and trees have died but the forest lives forever is the start of a poem I wrote years ago. It came to me while being still in the forest.

Much knowledge has been accumulated over the years in the forest and this knowledge can be gained from an open and deep connection with the forest.

Similarly, you can gain knowledge from an open and deep connection with whisky.

At one of my many tastings, we sampled 6 whiskies with a combined age statement of over 100 years. That was based on the youngest whisky in the glasses and you know those crafty master distillers could have added older whiskies into the mix but we will never know.

When you whisper whisky, you may be surprised at what comes to you.

Old people have wisdom accumulated from years of experience and lifelong learning.

So does old whisky.

Whisper whisky, gain knowledge, get wise!

The Value of a Good Cask

A good barrel will make a bad whisky better but a bad barrel will make a good whisky bad.

Be careful who you surround yourself with.

If you hang out with bad folk, they will make you bad. If you hang out with good folk, they will make you better.

Whisky folk are good folk, well most of the time, as long as they remember to respect the whisky.

However, it's more complex, obviously. It's kind of like nature and nurture. Which takes preference?

The nature of the whisky comes from the ingredients; the water, the malted barley, and the yeast, along with the physical bits...mash tuns, and the stills with various sized line arms complete with their own unique shapes and sizes and dents. These are critical to getting the brand expression just right.

If a bit of the still is worn out, before they replace it, they take a bunch of photos and make sure that the new parts have the same exact dents and patches of the old still.

It is that attention to detail that the Scots are famous for.

The nurture part is complex and comes from the care of the master distiller, and the maturation bit from the oak in the barrel that protects and shapes the whisky. I have learned that a cask in the middle of a storage warehouse can be glorious.

Each warehouse has a sweet spot where the best, most special select cask comes from.

Each cask is different and unique and imparts a different influence on the raw spirit.

I have had the honour, the privilege and the good fortune to taste the most glorious spirits, where nature and nurture are in perfect harmony.

I am beyond grateful for this experience.

Each warehouse has one or two exceptional casks which are worth a lot of money if they are even released for sale. If you can afford a bottle from those special cask releases, you will not be disappointed.

The emphasis on the influence of wood on the quality of whisky is rather modern, as distillers increasingly began to realize the importance of oak on the nurturing of fine whisky.

Whisky is the perfect vehicle to nurture us and adds to our nature when we whisper it.

I have seen wooden mash tuns converted into homes in the Findhorn Community in Scotland.

What a happy place that would be to live in.

Is the Glass Half Full or Half Empty?

It doesn't matter if there is whisky in it!

Years ago, when I was teaching in NW British Columbia, I heard a speaker who told an interesting story about the glass half full / half empty analogy. I seem to recall he was also a whisky man.

He didn't say what I thought he was going to say about the optimist sees the glass half full and the pessimist sees the glass as half empty. (Both by the way are 100% true.)

Instead, he said that the half full part represented the gifts that all people have and the half empty part represents the labels that we put on people.

The gifts are the positive productive skills that folk use to get things done.

The labels are the negative unproductive putdowns that people use to road block the productive bits.

He said the secret to a productive happy life is to focus more on people's gifts and less on their labels.

I like the fact that the half full part is the whisky and to add to his analogy, the whisky in the glass is the gift part and the half empty part is the room in the glass for more knowledge, supported by the whisky.

I have also seen somewhere that the half empty part represents the potential we have, or in teaching terms, that is the space needed for more knowledge to fill the glass.

Still though, I like the whisky part much more.

Toasting and Charring

The toasting and charring process in the casks brings out the colours and flavours of the whisky.

In life, we often have times when we get burned. Other times, we just get toasted.

In that process, our true character is developed.

Whisky does not have control over how or what cask it goes into, but it makes the best with its charred or toasted surroundings, and that ultimately controls its true character.

How we process these events also helps us to shape and show our true character.

How the whisky responds to the oak in the barrel, ultimately influences the quality of the character of the whisky.

How we respond to the whisky ultimately influences the quality and character of our life.

Respect the whisky, always!

John Clement

Variety is the Spice of Life

One of the things that brings me the most joy about tasting and whispering whisky is playing with variety.

The other night a good friend whom I call the "Captain" came over and we had a wee tasting.

We tried a 10 year old Tobermory, an unpeated Islay whisky, (yes there is such a thing) at 46.3 % ABV vs a 12 year old Tomatin, a highland whisky from their distillery just south of Inverness at 43% ABV.

In my whisperings, I pour about 1.5cl (15 ml) per tasting into my proper small tasting glasses that I got from visiting Highland Park in the Orkney Islands. This is the perfect amount for 3 sips.

Well guess what. The 12 year old Tomatin whisky was the winner. Then for fun, I threw in a Canadian whisky, Northern Harvest Blend that was declared World's best whisky for 2016 by Jim Murray, Whisky Bible, 2016. Also for fun, we added Lot 40, the Canadian Champion 2016 (Canadian Whisky Awards) into the mix. And low and behold, the Canadian Award's choice won out for me while the good Captain backed up Murray's pick of the World Champion because of the complex movement of the mouth feel.

I then and don't know why, circled back and had a wee taste of the Scotches again and low and behold, the 12 year old Tomatin regained it spot as my top pick for the evening. Why is that?

Truth be told, it doesn't matter.

As my Hapkido master, Doug, once said, "The why is yours, grasshopper, things are or they are not".

In this case and in many other whisperings, I have found that my taste can vary during the event, so again explaining why there are 31 whiskies on my top 10 list.

As they say in relationship status on Facebook, it's complicated. Whisky is that complicated.

Single malt Scotch whisky is the most diverse of all the spirits in the world just like Scots pine (*Pinus sylvestris*) are the most genetically diverse of all the pines.

Red pine (*Pinus resinosa*) has little genetic diversity, all trees look the same. This reminds me of vodka which all tastes very similar. (Or perhaps my taste buds are just not that finely calibrated yet.)

I am in constant search for the perfect whisky. I know you are out there. I will not rest until I find you.

So many whiskies, so little time. Oh the variety!

Aligning Your Energy With Your Itness...Light It Up

When you whisper whisky you can get lit up like a Christmas tree, especially when you are in the presence of good folk and good whisky.

I believe the secret of success is when you align your energy with your being. Perhaps yet another book.

For most of us, our beings stem from our doings.

When we go about life, doing this or doing that, with no clear direction for our being, life is random and all over the place.

I believe your doing should stem from, and support your being, and not the other way around.

Take the time to figure out who you are and allow your doings to support your being.

Once I developed my persona as the wandering whisky whisperer, my doings supported my being and I had alignment. I knew I had to travel, and research and share my learnings.

I also knew I had to try new whiskies. I also knew I had to always respect whisky or none of this would happen.

This alignment creates a flow and once you are in the flow, you will become ultimately successful.

So align your energy with your itness and taste more whisky. And have fun.

Alignment is the assignment!

Whisky Evaporates Faster Once the Bottle is Open!

Ever notice that the whisky in your bottles evaporates faster once those bottles are open.

The ones in the cupboard that aren't open, do not evaporate. Such an interesting phenomenon!

I have noticed that whiskies that taste better, have a higher than normal evaporation rate once opened. Why is that?

Wandering Whisky Whisperer fact: The better the whisky, the faster it evaporates.

Three Plus Six Equals Nine!

On the tours of distilleries, in the research, everywhere you look, you will be told that Scotch whisky is made with 3 ingredients... malted barley, yeast and water. In my humble opinion, while this is a fact, it represents a very narrow minded view.

My musings and wanderings tell me, there are actually 6 ingredients...the first 3, plus oak, time and the skill and attention to detail of the master distiller.

So where does nine come in to this equation. Well my friends, it has something to do with exponentials.

You will just have to whisper some more whisky to figure that one out! Cheers!

Access

Whisky tasting gives you access to folks that you would not normally have access to.

Whisky folk have a special spot for other whisky folk.

I have had audience with many people that social class wise, are above me and have developed close friendships with them because of our whisky appreciating bond.

To be invited to a whisky tasting is an honour and a privilege and should always be accepted.

Also I have found a deeper connection with the folks that I have tasted and whispered whisky with than with any other.

This is another benefit to whispering whisky. Be sure to thank the whisky for that. Cheers!

Meaning of Life

The forest has the answers to life's many questions, if you get quiet enough to hear the answers.

So does whisky.

As mentioned previously, "Whisky provides a pathway to the deeper meanings in life." (Matt, 2016.) (At one of my whisperings.)

This is why life makes so much sense after you have tasted a few whiskies.

The meaning of life was actually revealed to me during one of my tastings but I was too far gone to write it down. It had something to do with adventure, and growth and sharing and enjoyment and helping others, with caring, respect, understanding and fairness thrown in for good measure along with the word connection.

I do get glimpses, every now and then, into the meaning of life while whispering whisky, especially the enjoyment part. Cheers!

The Secret of Whispering Whisky

The secret of whispering whisky is not what you say about it, not what it says about you, but what you actually learn from it, just as the secret to being a good teacher is not what you teach your students, not what they learn from you, but what you learn from them.

Many teachings come together to influence the teacher. Many whisperings come together to influence the whisperer.

The student becomes the blend of all of their teachers, just like the whisperer becomes the blend of all the whiskies he or she has whispered.

Thank you, whisky, for teaching me so many things! Cheers!

Whispering Whisky is Beyond Fun!

Tasting whisky is fun!

Trying to figure out why you like or dislike the whisky is fun.

Trying to figure out which whisky you like best in that moment, is fun.

Trying to figure out the flavours you get and listening to the folk at the tastings having different experiences is fun.

The laughter that emerges after the first nosing of whisky at a tasting is an indicator of this.

The mindfulness and depth of connection that results from whispering whisky is beyond fun.

Writing this book and sharing my insights with you is beyond fun. I hope your reading of my book is beyond fun.

Make sure you have fun my friends. Cheers!

Not on Mondays

I do not whisper whisky on Mondays, unless it falls on a holiday or during the 12 days before Christmas. Then, I will take the following Tuesday off instead.

In fact, I never drink any alcohol on Mondays.

That is my secret as to why I am not an alcoholic.

My wife thinks I am in denial and I keep asking her if that is a river in Egypt.

But seriously folks, it is good to give your liver a rest. It also gives your habit a rest too.

Many times, if I have a proper tasting or whispering master class, or a party, say on a Thursday night, I may not have a taste, even on my porridge, from the Sunday to the Thursday before the tasting.

At least once a year, I will take a break from alcohol for a 3 week period and that, my friends, works for me.

Final Note

There is a parallel here and another point waiting to be made but it just hasn't been revealed to me as yet.

Is there something that has been whispered to you while whispering whisky that you need to share with me?

Would you? Could you?

Perhaps those and other insights into the complexities of life will reveal themselves with a bit more whisky as you wander down the path to becoming a true whisky whisperer.

If you have some stories, insights, or musing that you would like to share, please send them to me, I would love to hear about your journey. You can personal message me on Facebook.

Cheers to all whisky whisperers!

Slainte Mhath!

Recipes

Recipes the Wandering Whisky Whisperer finds interesting!

Eggnog

Eggnog with whisky instead of rum. Yum. yum. Yum. Simply substitute whisky for rum in your eggnog! Cheers!

Whisky as a Preservative

Once you open maple syrup, add a wee bit of whisky to it as a preservative...if fact some of the old timers used to save whisky bottles and used them to store their maple syrup in. Little known historical fact. Be sure to refrigerate once opened.

Porridge with Whisky

Add a table spoon of whisky to your morning porridge.

Add a table spoon of whisky to your morning porridge, along with a table spoon of real maple syrup and add two table spoons of 18% cream. Very yummy and a glorious way to start your day.

Proper Stewed Rhubarb.

A Wandering Whisky Whisperer Original

1 cup of rhubarb stalks, chopped into ¼ inch sections

¼ cup sliced frozen or fresh strawberries

¼ cup sugar

Combine all ingredients in a small pot and add just enough water to almost cover ingredients.

Add 1 oz of pure maple syrup and 1-2 shakes of Cinnamon powder

Simmer on med heat for 8 min max, less if you want the rhubarb al dente.

With a slotted spoon, remove rhubarb and strawberry mix from liquid. Let the fruit mixture cool in a bowl.

Now heat up residual liquid for 8 more minutes, simmering on med low to thicken into a syrup.

Cool liquid before adding to rhubarb strawberry mixture in the bowl. Very important to cool first so doesn't continue cooking rhubarb.

Lastly stir in 1 oz. of maple syrup and 1 oz. of whisky to the mixture in the bowl.

Serve over real vanilla ice cream, not the frozen dessert crap.

Yes, you can thank me!!!

Lanie's Infamous Secret Whisky Sour Recipe

My friend Lanie is a whisky woman. All women who appreciate good whisky are to be revered. Lanie is no exception.

We have had several whiskies and good times over the years. She was kind enough to share her secret whisky sour recipe with me and has given me permission to share it with you.

1 ½ shots whisky (shot and shots are the exact terms Lanie wrote down in her recipe)

½ shot of fresh squeezed lemon juice

¼ shot simple syrup

½ tsp maraschino cherry juice (liquid from bottle of maraschino cherries)

Put in a shaker. "Shake it baby!" (her words)

Pour into an ice filled glass.

Garnish with orange twist and a maraschino cherry with the stem.

Made with love. Thank you Lanie!

Your Own Special Whisky Recipes

Get Creative. Write down some of your favorites

Last Thoughts

Some folk call me the Whisky Pope.

I am not the Whisky Pope. I am the Wandering Whisky Whisperer. But if I was the Whisky Pope, I would change the communion wine to communion whisky and the church would be a happier place.

Let me end with a wee toast.

Fit a rare dram in your han...

For here's a toast to the nation with the most...

Glorious whisky in the lan....

Cheers!

Author's note

This book is written for those folks who have taken a wee wander down the whisky path on their journey to enlightenment, and to those just starting out. It offers the reader a new and proper way to experience mindfulness not normally, but sometimes, associated with tasting of such fine craftsmanship...this thing they call "uisge beatha", the water of life.

Obviously, I enjoy whisky and I hope you do too. I hope you enjoyed my musings and I look forward to hearing yours. Please always respect whisky and drink responsibly. Never drink and drive. Cheers!

The Wandering Whisky Whisperer.